WORKBOOK

Working for over
25 YEARS
WITH
Cambridge Assessment International Education

T0265938
Endorsed for learner support

Cambridge IGCSE™

Physics
Third Edition

Heather Kennett

HODDER EDUCATION

Cambridge International copyright material in this publication is reproduced under licence and remains the intellectual property of Cambridge Assessment International Education.

Exam-style questions [and sample answers] have been written by the authors. In examinations, the way marks are awarded may be different. References to assessment and/or assessment preparation are the publisher's interpretation of the syllabus requirements and may not fully reflect the approach of Cambridge Assessment International Education.

Third-party websites and resources referred to in this publication have not been endorsed by Cambridge Assessment International Education.

Every effort has been made to trace all copyright holders, but if any have been inadvertently overlooked the publishers will be pleased to make the necessary arrangements at the first opportunity.

Hachette UK's policy is to use papers that are natural, renewable and recyclable products and made from wood grown in well-managed forests and other controlled sources. The logging and manufacturing processes are expected to conform to the environmental regulations of the country of origin.

Orders: please contact Hachette UK Distribution, Hely Hutchinson Centre, Milton Road, Didcot, Oxfordshire, OX11 7HH. Telephone: +44 (0)1235 827827. Email education@hachette.co.uk Lines are open from 9 a.m. to 5 p.m., Monday to Friday. You can also order through our website: www.hoddereducation.com.

ISBN: 978 1398 310 575

© Heather Kennett 2021

First published in 2012
Second edition published in 2015
This third edition published in 2021 by
Hodder Education
An Hachette UK Company
Carmelite House, 50 Victoria Embankment, London EC4Y 0DZ

www.hoddereducation.com

Impression number   5 4 3 2 1

Year   2025 2024 2023 2022 2021

All rights reserved. Apart from any use permitted under UK copyright law, no part of this publication may be reproduced or transmitted in any form or by any means, electronic or mechanical, including photocopying and recording, or held within any information storage and retrieval system, without permission in writing from the publisher or under licence from the Copyright Licensing Agency Limited. Further details of such licences (for reprographic reproduction) may be obtained from the Copyright Licensing Agency Limited, www.cla.co.uk.

Cover photo © Zffoto – stock.adobe.com

Typeset in India by Integra Software Services Pvt. Ltd, Pondicherry

Printed in the UK

A catalogue record for this title is available from the British Library

Contents

Introduction

This new edition of the *Cambridge IGCSE Physics Workbook* is designed as a 'write-in' book for students to practise and test their knowledge and understanding of the content of the Cambridge IGCSE™ Physics course.

The sections are presented in the same order as those in the Student's Book, *Cambridge IGCSE Physics Fourth Edition*, and as in the Cambridge IGCSE™ Physics syllabus for examination from 2023. All questions have been marked as either Core or Supplement. At the end of every section, there are longer questions (Exam-style questions) which aim to introduce students to an examination format.

This Workbook should be used as an additional resource throughout the course alongside the Student's Book. The 'write-in' design is ideal for use in class by students or for homework.

This Workbook is designed to be used throughout the Cambridge IGCSE™ Physics course and, as such, its structure mirrors the Student's Book which also is part of this series. It should be used as extra practice for students to apply the theoretical concepts they have studied in the Student's Book. This Workbook covers what we feel is a representative selection of the content within the syllabus, although not every syllabus point has a specific question related to it.

1A Measurement, motion, mass, weight and density

Core

1 State the number of millimetres in each measurement.

    a 2 cm = ...

    b 0.4 cm = ...

    c 12 cm = ...

    d 0.5 m = ...

    e 1.4 m = ...

2 Convert each length to metres.

    a 1500 cm = ...

    b 150 cm = ...

    c 15 cm = ...

    d 1.5 cm = ...

3 Give each number as a power of ten with one figure before the decimal point.

    a 1000 = ...

    b 225 000 = ...

    c 650 = ...

    d 15 000 = ...

4 Give each number in full.

    a 10^4 = ...

    b 2.5×10^2 = ...

    c 1.5×10^6 = ...

    d 3.5×10^8 = ...

5 Give each decimal as a power of ten with one figure before the decimal point.

    a 0.001 = ...

    b 0.02 = ...

    c 0.0012 = ...

    d 0.0102 = ...

6 Give each fraction as **i** a power of ten, and **ii** a decimal.

    a $\dfrac{1}{100}$ **i** ....   **ii** ....

    b $\dfrac{2}{1000}$ **i** ....   **ii** ....

    c $\dfrac{3}{10\,000}$ **i** ....   **ii** ....

    d $\dfrac{4}{5000}$ **i** ....   **ii** ....

7 State each length in metres as a power of ten.

    a 5 mm = ...

    b 50 cm = ...

    c 5 km = ...

8 State the number of significant figures in each of the following measurements.

    a 1.53 m   number of significant figures = ...

    b 2.5×10^4 m   number of significant figures = ...

    c 0.016 m   number of significant figures = ...

9 Round 1.263 m to:

    a one significant figure ....

    b two significant figures ....

    c three significant figures. ....

10 Calculate the area of a rectangle of length 6.0 cm and width 4.5 cm, giving your answer to an appropriate number of significant figures.

    area = ...

11 Calculate the area of a right-angled triangle of sides 30 cm, 40 cm, 50 cm. Show your working.

    area = ...

12 Calculate how many bricks of size 5 cm × 2 cm × 3 cm a child would need to build a rectangular block of dimensions 15 cm × 15 cm × 10 cm. Show your working.

    number of bricks = ...

13 A student records the time taken for 20 complete swings of a pendulum to be 32 seconds. Calculate the period of the pendulum.

    period = ...

14 Four runners competing in a 1 km race finish the race with these times:

A 200 s	B 180 s	C 230 s	D 215 s

   a Identify the runner that finished first. ....

   b Identify the runner that finished last. ....

   c Calculate the average speed in m/s of runner **A**. Show your working.

      average speed of runner **A** = ...

15 A farmer goes for a drive in his tractor. The graph below shows how his distance from the farm varies with time. Distance OD = 270 m.

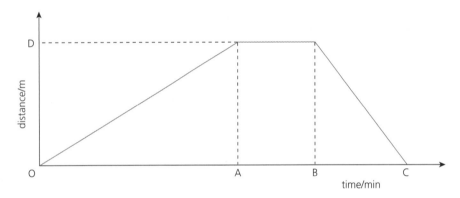

   a State how far the tractor travels in time interval:

     i OA ....

     ii AB ....

     iii BC ....

   b If BC = 3 minutes, calculate the speed in m/s of the tractor over time interval BC. Show your working.

      speed over time interval BC = ...

16 The Mars Rover Curiosity has a mass of 900 kg. Taking the gravitational field strength to be 9.8 N/kg on Earth and 3.7 N/kg on Mars, give the value of the weight of the Rover on:

   a Earth ....

   b Mars ....

17 Give the weight of the following masses. Take the gravitational field strength to be 9.8 N/kg.

a A mass of 15 g has a weight of ....

b A mass of 51 g has a weight of ....

c A mass of 0.3 kg has a weight of ....

d A mass of 3.1 kg has a weight of ....

18 On the Earth the acceleration of free fall is about 9.8 m/s². On the Moon the acceleration of free fall is about 1.6 m/s². A man weighs 784 N on the Earth. Give his:

a mass measured on the Moon

mass = ...

b weight measured on the Moon.

weight = ...

19 Circle the **correct** SI unit of density.

A g/cm²   **B** g/m³   **C** kg/m³   **D** kg/cm³

20 A block of ice has dimensions 2.0 cm × 2.0 cm × 2.0 cm and a mass of 7.36 g. Calculate the density of the ice. Show your working.

density = ...

21 A metal spanner has a mass of 200 g. It is lowered into a measuring cylinder of water until it is completely submerged. The original level of the water was 50 cm³ and the final level is 75 cm³. Calculate:

a the volume of the metal

volume = ...

b the density of the metal. Show your working.

density = ...

22 The density of air is $1.3\,kg/m^3$. Calculate the mass of air in a room of dimensions $3\,m \times 4\,m \times 2.5\,m$. Show your working.

mass of air = ...

Supplement

23 **a** A bus starts from rest and accelerates smoothly. After 10 s the bus reaches a speed of 8 m/s. Calculate the acceleration of the bus. Show your working.

acceleration = ...

b The bus continues to travel at 8 m/s and then decelerates smoothly as it approaches a bus stop. If the deceleration is $2\,m/s^2$, calculate the time over which the bus decelerates before it comes to rest. Show your working.

time = ...

24 A lorry is moving with a uniform acceleration of $1.5\,m/s^2$. At a certain time it is travelling at a speed of 6 m/s. Calculate the speed of the lorry 4 s later. Show your working.

speed = ...

25 The speed of a skier increases steadily from 8 m/s to 20 m/s in 60 s. Calculate:

a the average speed of the skier

average speed = ...

b the distance travelled by the skier in 60 s. Show your working.

distance = ...

c the acceleration of the skier. Show your working.

acceleration = ...

26 A girl rides her skateboard in a park. The graph below shows how her speed varies over a period of time.

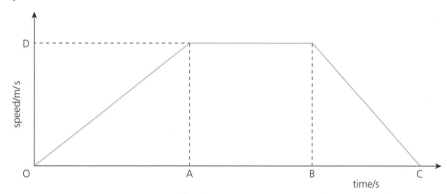

OD = 3 m/s, OA = 20 s, AB = 16 s, BC = 14 s. Calculate:

a the speed of the girl at time A

speed = ...

b her acceleration in time interval OA

acceleration = ...

c the distance she travels in time interval AB. Show your working.

distance in time interval AB = ...

1A MEASUREMENT, MOTION, MASS, WEIGHT AND DENSITY

d the distance she travels in time interval BC. Show your working.

distance in time interval BC = ...

27 An object falls from rest from the top of a high building. Ignore air resistance and take $g = 9.8\,\text{m/s}^2$.

a Calculate:

i the velocity of the object after 2 s

velocity = ...

ii the distance the object falls in 2 s. Show your working.

distance = ...

b Describe the shape and gradient of a graph of velocity against time for the object.

....

....

Exam-style questions

Core

1 The volume of a liquid in a measuring cylinder is 500 cm³. The liquid is poured into a square tank of internal dimensions 10 cm × 10 cm × 10 cm.

   a Calculate how full the container will be.

   Circle the correct answer.

   A completely full   **B** $\frac{1}{2}$ full   **C** $\frac{1}{4}$ full   **D** $\frac{1}{3}$ full   *[3]*

   b A brick of dimensions 4 cm × 3 cm × 5 cm is then lowered into the tank so that it is completely submerged. Calculate how far the liquid level rises.

   Circle the correct answer.

   A 0.6 cm   **B** 1.0 cm   **C** 1.5 cm   **D** 2 cm   *[3]*

   c A book has 120 pages and a thickness of 8.2 mm. The front and back cover of the book are each 0.5 mm thick. Calculate the thickness of each page.

   thickness = ...   *[3]*

   [Total: 9]

2 A woman goes for a walk to the local park. The graph below shows how her distance from home varies with time.

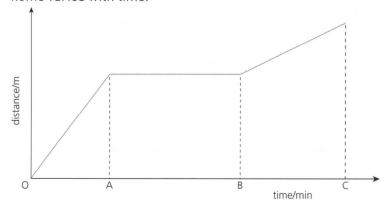

   Choose the correct answer from the box to the questions **a**, **b** and **c**.

   | zero | constant | increasing | decreasing |

a During the time interval OA,

    i the distance of the woman from O is .... *[1]*

    ii her speed is .... *[1]*

b During the time interval AB,

    i the distance of the woman from O is .... *[1]*

    ii her speed is .... *[1]*

c During the time interval BC,

    i the distance of the woman from O is .... *[1]*

    ii her speed is .... *[1]*

d Over which time interval is her speed greatest? .... *[1]*

[Total: 7]

Supplement

3 Oil with a volume of 50 cm3 has a mass of 46 g.

    a Calculate the density of the oil.

       density of oil = ... *[2]*

    b Vinegar of density 0.98 g/cm^3 is poured onto the oil; the two liquids do not mix. Explain why the vinegar sinks.

    ....

    .... *[2]*

[Total: 4]

4 Circle which of the following quantities is a vector.

    A density

    B velocity

    C time

    D mass

[Total: 1]

5 A man travels to work by car. He travels on local roads and on the motorway. The graph below shows how his speed varies with time during the journey.

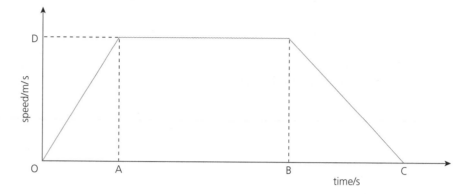

Take OD = 30 m/s, OA = 200 s, AB = 450 s and BC = 300 s.

a Calculate the acceleration of the car during OA.

acceleration = ... **[2]**

b Calculate the deceleration of the car during BC.

deceleration = ... **[2]**

c Calculate the average speed of the car in time interval BC.

average speed = ... **[2]**

d Calculate how far the car travels in time interval BC.

distance travelled = ... **[2]**

e Identify the time interval over which the average speed of the car is the highest.

.... **[1]**

[Total: 9]

1B Forces and momentum

Core

1 A box is subject to the forces shown in the diagram below.

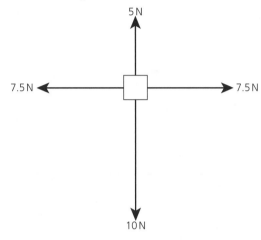

a Calculate the size and direction of the resultant force on the box. Show your working.

resultant force = ...

b State the size and direction of the extra force needed to reduce the resultant to zero. Show your working.

extra force = ...

2 Fill in the gaps in the following paragraph.

If there is no resultant .... on a body, it remains at .... or

continues to move at a .... speed in a .... line.

3 A skydiver of mass 80 kg jumps from a plane. Taking $g = 9.8 \, \text{m/s}^2$, determine:

a the weight of the skydiver

weight = ...

b the force due to gravity acting on the skydiver ....

c the resultant force on the skydiver when the air resistance is 300 N

resultant force = ...

d the value of the air resistance after the skydiver's parachute has opened and the skydiver has reached a constant speed. ....

4 A uniform metre ruler is balanced at its midpoint.

A weight, W, is hung from one end of the ruler and a load, L, is suspended at a distance X from the fulcrum, as shown.

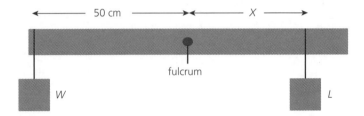

Calculate the distance X for the ruler to be balanced when:

a $L = 2W$

X = ...

b $L = 5W$

X = ...

c $L = 10W$

X = ...

5 Use the following words to fill in the gaps in the paragraph below.

clockwise	equals	forces	no	resultant
direction	equilibrium	moments	point	sum

When a number of parallel .... act on a body it will be in ....

when (i) the .... of the forces in one direction .... the sum

of the forces in the opposite .... and (ii) the sum of the ....

moments about any .... equals the sum of the anticlockwise ....

about the same point. This means that there is .... resultant force and no

.... turning effect when a system is in equilibrium.

6 A uniform board of length 90 cm is pivoted at a hinge at one end. It is kept level by an upward vertical force T applied at the opposite end. The weight of the board is 6 N.

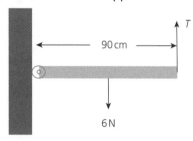

a Take moments about the hinge to calculate T when the board is level. Show your working.

T = ...

b The hinge exerts an upward force, F, on the board. Equate parallel forces to determine F when the board is level. Show your working.

F = ...

7 Explain how you would calculate the centre of gravity of a triangular-shaped piece of card using a pivot and a plumb line.

....

....

....

....

Supplement

8 A mass of 0.5 kg extends a spring by 10 cm. When an unknown mass, M, is hung on the spring, the extension is 15 cm. Calculate the value of M. Show your working.

M = ...

9 A force of 10 N extends a spring by 3.0 cm. Calculate the extension if a mass of 0.3 kg is attached to the end of the spring. Show your working.

Circle the **correct** answer.

A 0.9 mm   **B** 1.0 mm   **C** 3.0 mm   **D** 9.0 mm

10 Describe the difference between a scalar and a vector, and give an example of each.

....

....

....

11 Forces of 4 N and 3 N act at 90° to each other. Calculate the magnitude and direction of their resultant. Show your working.

magnitude of resultant = ...

direction of resultant = ...

12 A mass of 50 kg experiences a force F_1 to the right and a force F_2 to the left.

   a If $F_1 = 100\,N$ and $F_2 = 80\,N$, calculate:

      i the resultant force on the mass. Show your working.

         force = ...

      ii the acceleration of the mass. Show your working.

         acceleration = ...

   b If $F_1 = F_2 = 80\,N$, calculate:

      i the resultant force on the mass

         force = ...

      ii the acceleration of the mass.

         acceleration = ...

13 Calculate the resultant force that produces an acceleration of $4\,m/s^2$ in a mass of 15 kg.

      resultant force = ...

14 A girl whirls a ball on the end of a string in a vertical circle at a constant speed.

The velocity of the ball is changing.

   a Describe the difference between speed and velocity in terms of vectors and scalars.

     ....

     ....

   b Explain what causes the velocity of the ball to change.

     ....

     ....

   c When the girl whirls the ball faster and faster, the string breaks when the ball is at its lowest point. In which direction does the ball fly off?

     ....

15 The momentum of a body of mass m and velocity v is equal to $m \times v$. Calculate the momentum of a 5.0 kg trolley travelling at velocity:

   a 3 m/s

     momentum = ...

   b 40 m/s

     momentum = ...

16 An ice hockey puck moves in a straight line with a velocity of 5 m/s. It strikes a second identical puck, which is initially at rest, head-on. The second puck moves off in the same straight line as the first puck with a velocity of 4 m/s. Calculate the velocity of the first puck after the collision. Show your working.

velocity of first puck after collision = ...

17 A ball of mass 50 g is at rest before being struck by a bat. The collision between the bat and the ball lasts for 0.002 s, and the speed of the ball immediately after it leaves the bat is 20 m/s.

a Calculate the momentum of the ball just after it leaves the bat.

momentum of ball = ...

b Calculate the impulse acting on the ball during the collision.

impulse on ball = ...

c Calculate the steady force which the bat exerts on the ball during the collision.

force = ...

Exam-style questions

Core

1 The following measurements are obtained when a spring is stretched.

Stretching force/N	2	4	6	8	10
Extension/mm	1	2	3	4.2	6

a Sketch an extension–force graph for the spring. Plot extension on the y-axis and force on the x-axis. *[4]*

b Mark on your graph the region over which the plot is linear. *[1]*

c Calculate the gradient of your graph.

gradient = ... *[2]*

[Total: 7]

Supplement

2 a Calculate the spring constant of a spring which is stretched 10 cm by a mass of 2 kg.

spring constant = ... *[2]*

b The length of an unstretched spring is 20 cm. When a force of 0.5 N is applied to the spring it stretches to a total length of 30 cm. Assume the limit of proportionality is not exceeded.

Calculate the spring constant.

spring constant = ... *[2]*

c The spring in part **b** is now stretched to 40 cm. Assuming again that the limit of proportionality is not exceeded, calculate the extension and the force applied.

i extension of spring = ... *[1]*

ii force applied = ... *[1]*

[Total: 6]

3 A trestle consists of a plank resting on two supports which exert upward forces of P and Q, as shown. The weight of the plank is 80 N (shown acting through its centre).

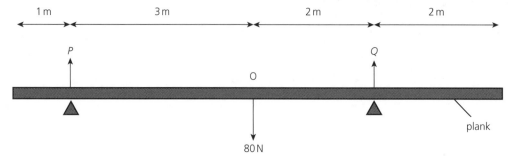

a Take moments about the left support to calculate the value of Q.

Q = ... **[2]**

b Equate parallel forces to calculate the value of P.

P = ... **[2]**

[Total: 4]

4 A golfer strikes a stationary golf ball of mass 45 g with a golf club. The golf club is in contact with the ball for a time of 0.001 s and exerts a force of 1350 N on it.

a Calculate the acceleration of the ball during the time it is in contact with the club.

acceleration = ... **[2]**

b Calculate the velocity of the ball just after it is struck.

velocity = ... **[2]**

c Give **two** ways by which the velocity of the golf ball could be increased.

1 ....

2 .... **[2]**

[Total: 6]

1C Energy, work, power and pressure

Core

1 State how energy is transferred from:

a a steam turbine

  ....

b a generator.

  ....

2 State the stores between which energy is transferred:

a when a stone falls from a cliff

  ....   →   ....

b when coal is burnt to heat water in a boiler.

  ....   →   ....

3 A dog pulls a sledge across a flat snowfield for a distance of 500 m. The dog exerts a steady force of 120 N on the sledge. Calculate the work done. Show your working.

  work done = ...

4 When a force of 150 N is applied to a piston X, it moves through a distance of 10 cm in the direction of the force.

a Calculate the work done by piston X. Show your working.

  work done = ...

b Another piston, Y, does the same amount of work as piston X in part **a**. The force on piston Y is 750 N. Calculate how far piston Y moves in the direction of the force. Show your working.

distance = ...

5 Explain what is meant by 'renewable' and 'non-renewable' energy resources. Give **two** examples of each type of resource.

a renewable energy resources:

....

example: ....

example: ....

b non-renewable energy resources:

....

example: ....

example: ....

6 A lift raises a load of 9000 N to a height of 20 m in 15 seconds.

a Calculate the work done. Show your working.

work done = ...

b Calculate the power of the lift. Show your working.

power = ...

7 Calculate the pressure on a surface when a force of 40 N acts on an area of:

a 4.0 m^2 **b** 0.5 m^2

pressure = ...   pressure = ...

c $0.1\,m^2$ **d** $4.0\,cm^2$

pressure = ...   pressure = ...

8 Calculate the force on each area if a pressure of $20\,Pa$ acts on it.

a area = $2\,m^2$ **b** area = $1\,m^2$

force = ...   force = ...

c area = $1\,cm^2$ **d** area = $0.1\,cm^2$

force = ...   force = ...

9 A boy decides to go for a walk in some soft snow. Which of the following footwear should he choose so that he sinks into the snow the least distance?

Circle the **correct** answer.

A boots of contact area $0.7\,m^2$

B snow shoes of contact area $1.5\,m^2$

C skis of contact area $1.0\,m^2$

D ice skates of contact area $5 \times 10^{-4}\,m^2$

10 In a hydraulic jack a force of $100\,N$ is applied to a piston of surface area $4.0\,cm^2$.

a Calculate the pressure applied to the fluid by the piston. Show your working.

pressure = ...

b The fluid transmits the pressure to a larger piston of surface area $20\,cm^2$.

    i Calculate the force exerted by the larger piston. Show your working.

force = ...

    ii Calculate the weight that can be lifted by the jack.

weight = ...

Supplement

11 Calculate the kinetic energy of a mass of $5\,kg$ travelling at a velocity of:

   a $6\,m/s$ **b** $12\,m/s$

Show your working.

E_k = ... E_k = ...

12 Taking $g = 9.8\,N/kg$, calculate the gravitational potential energy gained by a mass of $5.1\,kg$ raised to a height of:

   a $10\,m$ above the ground   **b** $25\,m$ above the ground

Show your working.

ΔE_p = ... ΔE_p = ...

13 Calculate the velocity of a truck with kinetic energy $2.7\,kJ$ and mass $600\,kg$. Show your working.

velocity = ...

14 A box of mass 3 kg is dropped from a height of 5.1 m.

    a Neglecting air resistance, calculate and show your working for:

        i the gravitational potential energy of the box before it is dropped

        E_p = ...

        ii the kinetic energy of the box just before it reaches the ground

        E_k = ...

        iii the velocity of the box just before it strikes the ground.

        velocity = ...

    b Explain what happens to the kinetic energy of the box when it strikes the ground.

    ....

15 Water flows over a dam wall at a rate of 2000 kg/s. The dam wall is 10 m high. Calculate how much power can be generated by the falling water if 90% of its potential energy can be harnessed to produce electricity. Show your working.

    power generated = ...

16 An electrical appliance has a power input of 500 W.

    a The appliance transfers energy at a rate of 350 W. Calculate its efficiency. Show your working.

        efficiency = ...

    b Explain what happens to the 'lost' energy'.

    ....

17 A swimming pool contains water at a depth of 3.05 m. Calculate the difference in water pressure between the top and bottom of the pool. Show your working. Take the density of water to be 1.0×10^3 kg/m³.

pressure difference = ...

Exam-style questions

Core

1 Two identical ropes are attached to either end of a beam of weight 600 N.

   a Calculate the tension (force), T, in each rope when they support the beam horizontally above the ground.

     T = ...   *[1]*

   b Calculate the work done if the beam is raised vertically by 1.5 m.

     work done = ...   *[2]*

   c Calculate the power needed to raise the beam 1.5 m in 3 seconds.

     power = ...   *[2]*

   [Total: 5]

2 Give **two** advantages and **two** disadvantages of using the following as sources of energy for electricity generation.

   a solar energy

     i advantages: ....

     .... *[2]*

     ii disadvantages: ....

     .... *[2]*

b nuclear energy

   i advantages: ....

   .... *[2]*

   ii disadvantages: ....

   .... *[2]*

c wind energy

   i advantages: ....

   .... *[2]*

   ii disadvantages: ....

   .... *[2]*

d gas-fired power stations

   i advantages: ....

   .... *[2]*

   ii disadvantages: ....

   .... *[2]*

[Total: 16]

Supplement

3 a A cyclist freewheels (without pedalling) from rest down a hill. Explain the energy transfers which occur during her descent.

   ....

   ....

   ....

   .... *[3]*

b On reaching level ground the cyclist pedals along a straight road with a constant speed of 12 m/s. She experiences a resistive force, F, of 5 N, which acts in the opposite direction to that in which she is travelling. Calculate the work she does against F in 1 second.

   work done = ... *[2]*

c   *F* then increases to 7.5 N. Calculate the cyclist's new speed, if she maintains the same rate of working as in part **b**.

i   rate of working = ...   *[1]*

ii   speed = ...   *[2]*

[Total: 8]

4   A ball of mass 60 g is projected vertically upwards from ground level with an initial velocity of 30 m/s. Neglect the effect of air resistance on the motion of the ball. Determine:

a   the initial kinetic energy of the ball

initial E_k = ...   *[3]*

b   the initial potential energy of the ball

initial E_p = ...   *[1]*

c   the potential energy of the ball when it reaches its greatest height

E_p at greatest height = ...   *[1]*

d   the greatest height reached by the ball.

greatest height = ...   *[2]*

[Total: 7]

2 Thermal physics

Core

1 Identify the state of matter – solid, liquid or gas – in which the molecules are:

   a least densely packed ....

   b ordered in a regular pattern ....

   c moving at high speed over large distances ....

   d moving about over small distances ....

   e vibrating towards and away from a fixed point. ....

2 Identify the state of matter – solid, liquid or gas – which:

   a is highly compressible ....

   b has a definite shape ....

   c flows easily. ....

3 Some smoke particles are allowed to drift into a glass box which contains air. The box is then sealed and illuminated. The random motion of the smoke particles in the box is viewed through a microscope.

   a Explain the random motion of the smoke particles with reference to the size, number and motion of air molecules.

   ....

   ....

   ....

   ....

   ....

   ....

   b The box is heated so that the temperature of the air it contains rises. Explain how the rise in temperature affects the motion of the smoke particles.

   ....

4 Use the kinetic particle model to explain how a gas exerts a pressure on the walls of its container; give **two** key facts.

....

....

5 The air in a closed container is heated.

    a State how the air pressure inside the container changes when the temperature rises.

    ....

    b Explain your answer to **a** in terms of the kinetic particle model. State the effect on the motion of the molecules, their interaction with the walls and the resulting effect.

    ....

    ....

    ....

6 The volume of a gas remains constant when it is heated in a closed container.

Circle the statement that is **false**.

    A The pressure of the gas will increase.

    B The average kinetic energy of the molecules will decrease.

    C The molecules will hit the walls of the container more often.

    D The molecules will move in all directions.

7 A fixed mass of gas is heated in a container which maintains the gas at constant pressure.

Circle the statement that is **true**.

    A The volume of the gas will decrease.

    B The volume of the gas will increase.

    C The number of molecules will decrease.

    D The molecules will hit the walls of the container more often.

8 On the Celsius scale of temperature, state the value of:

    a the melting temperature of ice ....

    b the boiling temperature of water at normal pressure. ....

9 A temperature has a different value in degrees Celsius than it has in Kelvin.

   a State the relationship between a temperature T (in K) to its value θ (in °C).

   ....

   b Convert a temperature of 30°C to a value in Kelvin. ....

   c Convert a temperature of 150 K to a value in °C. ....

10 State the meaning, with reference to states of matter, of:

   a melting temperature

   ....

   ....

   b boiling temperature

   ....

   ....

11 Which of these materials would you choose to use for the following applications?

A wood	B glass	C steel	D fibreglass	E cardboard

   a a cooking pot ....

   b the handle of a saucepan ....

   c a container for a takeaway pizza ....

   d roof insulation for a house ....

12 Describe an experiment, using metal and wood rods (of the same dimensions), to show the difference between a good conductor and a bad conductor of thermal energy.

   ....

   ....

   ....

   ....

   ....

   ....

13 Explain why a metal object at room temperature feels cold compared to a plastic object at the same temperature.

....

....

14 Circle the following statement that is **incorrect**.

A hot air rises

B cold air sinks

C cold air is less dense

D hot air is less dense

15 Circle the **correct** answer below.

Convection takes place in:

A only solids

B only liquids

C only gases

D gases and liquids

16 These are features of a vacuum flask:

A double-walled glass container

B vacuum

C silvered glass surfaces

D metal container

State which of the above features reduces the transfer of thermal energy by:

a both conduction and convection ....

b radiation. ....

17 Use the following words to fill in the gaps in the paragraph below.

electromagnetic	fluid	lower	temperature
energy	higher	radiation	thermal

Conduction is the transfer of .... energy through matter from places of

.... temperature to places of .... temperature without

movement of the matter as a whole. Convection is the transfer of thermal ....

through a fluid from places of higher .... to places of lower temperature by

movement of the .... In ...., thermal energy is transferred from

one place to another by means of .... waves; no medium is required.

Supplement

18 A fixed mass of gas occupies a volume of 200 cm³ at a temperature of 27°C and a pressure of 1 atmosphere. If the temperature is kept constant, calculate and show your working for:

a the volume occupied when the pressure is halved

volume = ...

b the volume occupied when the pressure is doubled.

volume = ...

19 A fixed mass of gas occupies a volume of 100 cm³ at a pressure of 2×10^5 Pa. The temperature of the gas is kept constant. Calculate and show your working for:

a the pressure when the volume is doubled

pressure = ...

b the pressure when the volume is halved.

pressure = ...

20 The following amounts of energy in joules are needed to raise the temperature of 1 kg of different materials, **A**, **B**, **C** and **D**, by 1°C.

A 4000   **B** 960   **C** 450   **D** 390

a Which material has the highest, specific heat capacity? ....

b Define 'specific heat capacity'.

....

....

21 Passive solar houses are energy-efficient buildings designed to trap and store sunlight in the winter. They often include an internal feature of a brick or concrete wall close to a window. Explain the purpose of the wall and why brick or concrete is used.

....

....

22 A cup of coffee is heated in a microwave oven. The mass of the liquid is 70 g and its specific heat capacity is 4000 J/kg °C. Calculate the energy needed to raise the temperature of the coffee by 60°C. Show your working.

energy required = ...

23 A 150 W immersion heater is used to heat 500 g of water in an insulated container. It takes 9 minutes and 20 seconds to raise the temperature of the water from 20°C to 60°C. Calculate the specific heat capacity of the water. Show your working.

specific heat capacity of water = ...

24 For **a** and **b**, circle **one** statement that is incorrect.

  a During melting:

    A the temperature remains constant

    B the average kinetic energy of the particles increases

    C the distance between particles increases

    D thermal energy is absorbed.

  b During solidification:

    A the temperature of the material increases

    B the average kinetic energy of the molecules remains constant

    C the distance between particles decreases

    D thermal energy is given out.

25 This question is about evaporation.

a Use the kinetic particle model to explain why a liquid cools when evaporation from its surface occurs. Compare the speed and kinetic energy of particles escaping from, or remaining in, the liquid.

....

....

....

....

b Identify **two** factors that increase the rate of evaporation of a liquid.

1 ....   **2** ....

26 This question is about radiation.

a Two copper blocks are immersed in hot water until they reach the same temperature. The blocks are identical except that block A is painted black and block B has a shiny surface. When the two blocks are removed from the hot water and placed well separated on a wooden board, which block cools faster? Give a reason for your answer.

....

....

b Explain why:

i shiny aluminium foil is sometimes placed behind central heating radiators

....

ii it is cooler to wear a white t-shirt than a black one on a hot day.

....

Exam-style questions

Core

1 There are three states of matter: solid, liquid and gas.

a Explain how the temperature of a liquid changes when it **i** boils and **ii** solidifies.

i ....

ii ....   *[2]*

b State how the separation of the particles in a liquid changes when it **i** boils and **ii** solidifies.

i ....

ii .... *[2]*

c Explain what happens to a liquid during evaporation.

....

....

.... *[3]*

[Total: 7]

2 Air is a poor conductor of thermal energy but a good insulator.

a Give **three** everyday examples of the use of trapped air to provide insulation.

1 ....

2 ....

3 .... *[3]*

b Explain why using trapped air for insulation is more effective than using air alone.

.... *[1]*

[Total: 4]

Supplement

3 Describe an experiment to measure the specific heat capacity of an aluminium block.

....

....

....

....

....

....

....

....

....

[Total: 8]

4 This question is about conduction.

   a Use the kinetic particle model to explain how thermal energy is transferred through a solid object.

....

....

....

.... *[4]*

   b Explain why poor electrical conductors are also poor thermal conductors.

....

.... *[1]*

[Total: 5]

3 Waves

Core

1 Explain the difference between a transverse wave and a longitudinal wave.

....

....

2 The diagram below shows a displacement–distance graph of a wave at a particular instant.

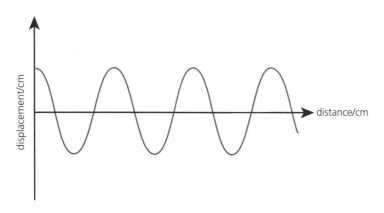

Label a crest, a trough, the amplitude and the wavelength.

3 A wave crest passes a particular point every $\frac{1}{10}$th of a second. Calculate the frequency of the wave.

frequency = ...

4 A heart rate is timed at 72 beats a minute. Calculate the frequency of the heart rate. Show your working.

frequency = ...

5 The lines drawn below represent the crests of water waves in a ripple tank.

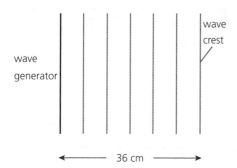

The distance occupied by six waves is 36 cm. The frequency of the wave generator is 5 Hz. Calculate:

a the wavelength of the waves

b the frequency of the waves

wavelength = ...

frequency = ...

c the speed of the waves

d the time taken to travel 9.6 m.

wave speed = ...

time taken = ...

6 When a water wave enters shallower water, its speed changes from 10 m/s to 5 m/s. State the changes that occur to:

a the frequency ....

b the wavelength ....

c the direction of travel. ....

7 Use some of the following words to fill in the gaps in the paragraph below.

after	beam	before	direction	diverging	faster
lines	narrower	slower	ray	wider	

Light travels in straight .... In diagrams a ray is used to represent the

.... in which the light is travelling. A .... is drawn as a straight

line with an arrow on it. A .... of light consists of many rays which may be

parallel, ... or converging. A diverging beam spreads out, while a converging

beam becomes .... Light travels much .... than sound. We see a

lightning flash .... we hear the corresponding sound of the thunder.

8 Sketch light rays in the following types of beam:

a parallel **b** converging **c** diverging

9 A ray of light is reflected from a plane mirror, as shown.

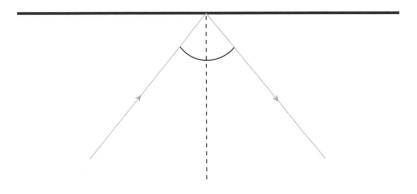

a Label the plane mirror, the incident ray, the normal to the mirror, the reflected ray, the angle of incidence, i, and the angle of reflection, r.

b Use a protractor to determine:

i the angle of incidence, i = ...

ii the angle of reflection, r = ...

10 In each of the following diagrams:

    a sketch the normal to the plane mirror at the point the light ray strikes it

    b mark on the angle of incidence, *i*

    c sketch the reflected ray.

i

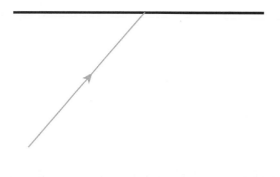

ii

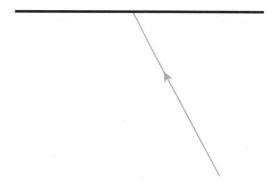

iii

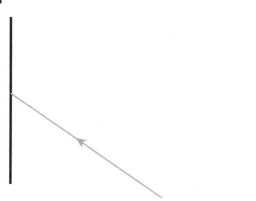

iv

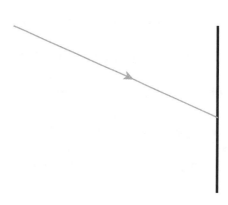

v

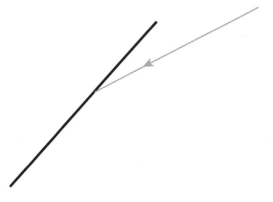

vi

11 Two plane mirrors, AB and BC, are set up at right angles. ON is a normal to the mirror AB. A ray of light XO is incident on AB, as shown. It is reflected at mirror AB and then travels on to mirror BC, where it is again reflected.

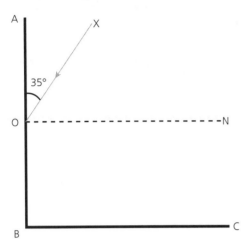

a On the diagram, continue the ray XO to show the path it takes after reflection at each mirror.

b Determine the following values:

   i   angle of incidence at mirror AB = ...

   ii angle of reflection at mirror AB = ...

   iii angle of incidence at mirror BC = ...

   iv angle of reflection at mirror BC = ...

c Comment on the path of the reflected ray.

   ....

12 Circle the statement that is **false**.

   The image in a plane mirror is:

   A as far behind the mirror as the object is in front

   B larger than the object

   C virtual

   D upright.

13 Explain what is meant by a 'real' image. Give **two** key facts.

   ....

   ....

14 A girl is standing 0.5 m away from a plane mirror.

   a State how far away the girl is from her image in the mirror.

   distance from image = ...

   b Determine how far the girl must walk away from the mirror to be 3 m from her image.

   distance walked = ...

15 Describe what happens to a ray of light when it passes from air into a different material, such as glass or water.

   ....

   ....

16 Use the following words to fill in the gaps in the paragraph below.

away from	denser	normal	normally	optically	towards

   A ray of light is bent .... the normal when it enters an optically

   .... medium and .... the .... when it enters an

   .... less dense medium. When a ray strikes a surface ...., it is

   not refracted.

17 Circle the diagram which shows the ray of light refracted correctly.

   A

   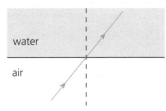

   C

   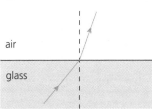

   B

   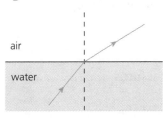

   D

   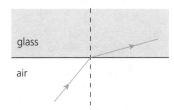

18 When sunlight falls on a triangular glass prism, a band of colours (called a spectrum) is obtained.

   a Explain why this happens.

   ....

   ....

b Give the colours of the spectrum, in order of diminishing wavelength.

....

....

c Sketch the path of red and blue rays of light through the prism.

prism

19 Explain what is meant by 'total internal reflection' in terms of optical density, angle of incidence and critical angle.

....

....

....

20 Use the following words to fill in the gaps in the paragraph below.

| centre | F | focus | parallel | principal |
| refracted | thin | top | two | undeviated |

The image of an object formed by a .... converging lens can be found by drawing

.... of the following three rays:

• a ray from the top of the object .... to the principal axis of the lens which is

refracted through the principal ....

• a ray from the top of the object which passes .... through the optical

............................., C, of the lens

• a ray from the .... of the object through the principal focus, ...,

which is .... parallel to the .... axis of the lens.

21 An object is placed 15 cm from a converging lens of focal length 12 cm. Circle the **correct** description of the image.

A virtual, erect and larger

C real, inverted and the same size

B real, inverted and smaller

D real, inverted and larger

22 Explain how you could estimate the focal length of a converging lens.

....

....

23 A small lamp is placed at the principal focus of a converging lens. Identify the type of beam that

will be produced by the lens. ....

24 This question is about electromagnetic waves.

a Identify **four** properties that are common to all types of electromagnetic wave.

1 ....   **3** ....

2 ....   **4** ....

b Identify the type of electromagnetic wave used in:

i intruder alarms ....

ii cooking ....

iii detecting broken bones ....

iv remote controllers for TVs ....

v luggage screening at airports ....

vi mobile phones. ....

25 Identify which of the waves in the box below:

infrared light	blue light	X-rays	radio waves

a has the longest wavelength ....

b has the highest frequency ....

c is the most penetrating. ....

26 The speed of light in air is 3×10^8 m/s.

a Calculate the wavelength in air of violet light of frequency 7.5×10^{14} Hz. Show your working.

wavelength of violet light = ...

b Calculate the wavelength in air of red light of frequency 4.3×10^{14} Hz. Show your working.

wavelength of red light = ...

27 Identify the risk associated with and a precaution that can be taken when:

a using a mobile phone

risk: ....

precaution: ....

b taking a medical X-ray photograph.

risk: ....

precaution: ....

Supplement

28 Two light rays, OP and OQ, from a point object, O, are incident on a plane mirror.

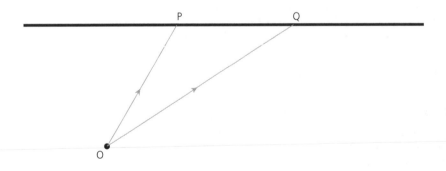

a Sketch the normals to the mirror at the points of contact of the rays, and sketch the reflected rays. Use dotted lines to extend the reflected rays behind the mirror to locate and mark the position of the image, I.

b State the characteristics of the image.

....

....

....

c Comment on the distance of the image from the mirror.

....

29 Light travels at a speed of 3×10^8 m/s in air. Calculate and show your working for the speed of light in:

a glass of refractive index $\frac{3}{2}$

b water of refractive index $\frac{4}{3}$

speed of light in glass = ...

speed of light in water = ...

30 a An optical fibre is made of glass of refractive index 1.5. Sketch the path of a ray that undergoes total internal reflection in the optical fibre.

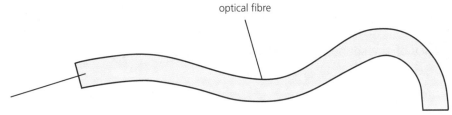

optical fibre

b Are analogue or digital signals preferred for transmitting data in optical fibres?

....

31 Water has a critical angle of 49°.

a Sketch what happens to a ray of light incident on a water/air boundary at the critical angle.

b Calculate the refractive index of water using the equation $n = \dfrac{1}{\sin c}$. Show your working.

refractive index = ...

32 The critical angle for light travelling in a certain liquid is 52°. Calculate the refractive index of the liquid. Show your working.

refractive index = ...

33 a Circle which **one** of the following does not change when red light enters glass from air.

    A speed   **B** frequency   **C** wavelength   **D** angle ray makes to the normal

b Define the term 'monochromatic'.

....

34 A sound wave has a wavelength of 20 cm and a speed of 330 m/s.

a Calculate the frequency of the sound wave. Show your working.

frequency of wave = ...

b State a typical value for a sound wave of:

    i audible frequency ....

    ii ultrasonic frequency ....

35 Calculate how far away the storm is when 8 seconds elapse between a lightning flash and the clap of thunder. Show your working. (Sound travels at 330 m/s.)

distance of storm = ...

Exam-style questions

Core

1 Circle the statement that is **true**.

The image in a plane mirror is:

A as far behind the mirror as the object is in front

B smaller than the object

C real

D inverted.

[Total: 1]

2 Right-angled glass prisms A and B are made of glass that has a critical angle of 42°.

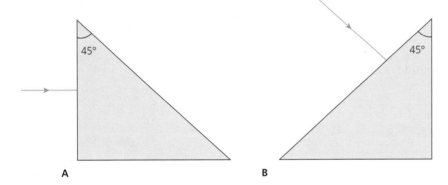

A   B

  a Sketch the path of the rays through prisms A and B.   *[5]*

  b Identify an instrument for each that would use prisms in orientation:

    i A ....   *[1]*   **ii** B ....   *[1]*

[Total: 7]

3 **a** Sketch a half-size ray diagram to locate the image of an object placed on the principal axis, 10 cm from the optical centre of a thin converging lens of focal length 5 cm. Mark the position of the principal foci, optical centre, object and image on your diagram.   *[4]*

  b Determine how far from the lens the image is formed. ....   *[1]*

  c Determine the size of the image compared to that of the object. ....   *[1]*

[Total: 6]

4 Sailors on a ship hear the echo of their foghorn from a cliff, 3 s after it sounded.

   a Calculate the distance of the ship from the cliff. Assume that the sound travels at 330 m/s.

   distance of ship from cliff = ... **[3]**

   b The ship moves 165 m towards the cliff. Calculate the time interval between the sounding of the foghorn and its echo reaching the ship.

   time interval = ... **[3]**

   [Total: 6]

5 The speed of sound in air is 330 m/s.

   a Calculate how far a sound travels in air in a time interval of 3 ms.

   distance = ... **[2]**

   b Sound takes 0.15 ms to travel down an iron bar of length 0.9 m. Calculate the speed of sound in iron.

   speed = ... **[2]**

   c Describe an experiment to determine the speed of sound in air.

   ....

   ....

   ....

   ....

   .... **[4]**

   [Total: 8]

6 Identify the following waves as transverse or longitudinal:

a sound	....	**[1]**	**d** water	....	**[1]**
b red light	....	**[1]**	**e** radio	....	**[1]**
c X-rays	....	**[1]**	**f** ultrasonic	....	**[1]**

   [Total: 6]

Supplement

7 In a ripple tank experiment, straight waves are incident on a barrier with a gap.

On the diagram, sketch the wave pattern after the waves pass through:

a a gap that is narrow compared with the wavelength [3]

b a gap that is wide compared with the wavelength. [3]

[Total: 6]

8 A ray of light is incident at 45° on mirror M1. It is reflected and strikes a second mirror M2. M2 lies parallel to M1.

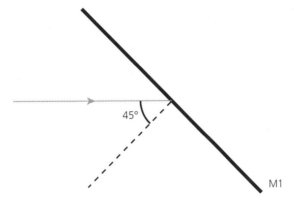

a Sketch the path of the ray on the diagram to show how it is reflected from M2. [4]

b For your diagram in part **a**, determine:

i the angle of reflection at M2 = ... [1]

ii the type of instrument in which this arrangement of mirrors could be used.

.... [1]

iii Give an advantage the instrument would have if the separation of M1 and M2 was increased.

.... [1]

[Total: 7]

9 A ray of light is incident as shown on a parallel-sided glass block.

a Sketch the normal to the block at the point X. Sketch the path of the ray through the block and out the other side. Mark the angle of incidence, *i*, and the angle of refraction, *r*, on the diagram. [4]

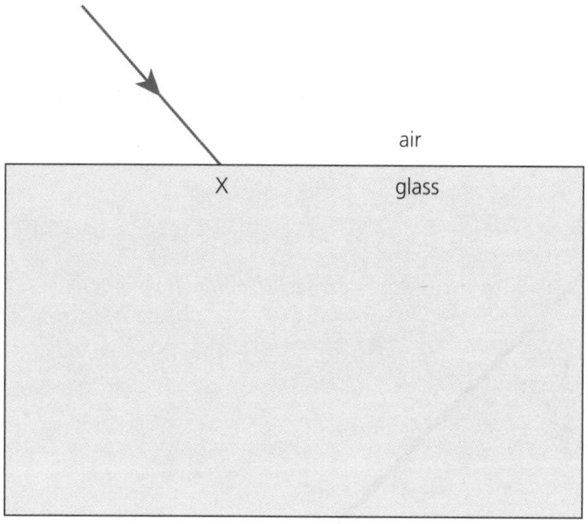

air

X   glass

b Comment on the direction of the ray when it leaves the block.

....

.... [2]

c The angle of incidence is 30° and the refractive index of the glass block is 1.5. Calculate the angle of refraction at X.

angle of refraction = ... [4]

[Total: 10]

10 A small insect is viewed with a magnifying glass of focal length 4 cm. The insect is 2 cm from the lens.

a Sketch a ray diagram to locate the image of the insect. **[5]**

b Determine the distance, v, of the image from the lens.

v = ... **[1]**

c State the characteristics of the image.

....

....

.... **[3]**

[Total: 9]

11 A laser directs a narrow beam of light along a radius of a semicircular glass block of refractive index 1.5, as shown.

a Explain why the light is not refracted when it enters the glass at P.

.... **[1]**

b Sketch the normal at Q and determine the angle of incidence, i, at Q.

i = ... **[2]**

c Calculate the critical angle for the glass, using the equation $\sin c = \dfrac{\sin 90°}{n}$ where n is the refractive index of the glass.

critical angle = ... **[3]**

d Sketch the path of the ray after it reaches Q. **[2]**

[Total: 8]

12 Two musical notes X and Y are represented by the following waveforms.

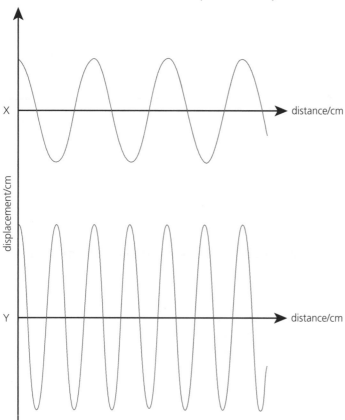

a Identify which of X and Y has:

i the higher frequency .... **[1]**

ii the longer wavelength .... **[1]**

iii the louder sound. .... **[1]**

b Explain what is meant by 'compression' and 'rarefaction' in a sound wave.

....

....

....

....

.... *[3]*

c The distance between two consecutive compressions of a sound a wave is 0.85 m and the speed of sound in the medium is 340 m/s. Calculate the:

i wavelength = ... *[1]*

ii frequency = ... *[2]*

[Total: 9]

13 In a medical ultrasound imaging system, ultrasonic pulses from a transducer placed on the patient's skin are reflected from an internal organ. The pulses travel at 1400 m/s through the body. There is a 40 µs time delay between the transmitted and reflected pulse arriving back at the transducer.

a Calculate the depth inside the body of the reflecting organ.

depth = ... *[4]*

b State the relationship between the speed, frequency and wavelength of a wave.

.... *[1]*

c Calculate the wavelength of ultrasonic waves of frequency 10^6 Hz travelling at 1400 m/s.

wavelength = ... *[2]*

[Total: 7]

4A Electricity and magnetism

Core

1 Fill in the gaps in the sentences below using the following words.

attract	permanent	pole	repel	soft	south	steel

A bar magnet has a north .... and a .... pole. Like poles

.... and unlike poles .... Magnetic materials, such as iron, which

are easily magnetised and easily lose their magnetism, are termed ....

Hard magnetic materials, such as ...., are used in .... magnets.

2 The magnetic field around a bar magnet can be visualised as magnetic field lines.

a Describe an experiment which can be used to determine the pattern of field lines around a bar magnet using a plotting compass, paper and pencil.

....

....

....

....

....

....

....

....

b Sketch the pattern of magnetic field lines around a bar magnet, marking on the N and S poles.

Cambridge IGCSE™ Physics Workbook Third Edition

3 Circle **two** of the following materials that a magnet would not attract.

   A iron   **C** steel

   B aluminium   **D** copper

4 Give **two** ways a magnet can be made from a steel bar.

   1 ....

   2 ....

5 When the following charges are brought close together, do they attract or repel each other?

   a +   + ...

   b –   –   ....

   c +   –   ....

6 Explain, in terms of electron movement, what happens when a Perspex rod is charged positively by being rubbed with a cloth.

   ....

   ....

7 Sketch a circuit diagram containing a battery, an ammeter and two lamps which are connected

   a in series   **b** in parallel.

   On each diagram place the ammeter so that it measures the total current from the battery.

   a series circuit   **b** parallel circuit

8 Explain the difference between alternating current (a.c.) and direct current (d.c.).

   ....

   ....

9 A certain type of lamp reaches its full brightness when a 12 V battery is connected across it. Comment on the brightness of such lamps when:

a a 12 V battery is connected across two of the lamps connected in series

....

b a 12 V battery is connected across two of the lamps connected in parallel

....

c a 6 V battery is connected across two of the lamps connected in parallel.

....

10 This question is about electrical resistance.

a Give the relationship between potential difference, current and resistance.

....

b Explain how the circuit is connected, what measurements are taken and what calculations are made in an experiment to determine the resistance of a wire using a variable resistor, a voltmeter and an ammeter.

....

....

....

....

....

....

11 Calculate the p.d. across a 4 Ω resistor when there is a current of 0.3 A in it. Show your working.

p.d. across resistor = ...

12 There is a current of 2 A in a lamp when a p.d. of 12 V is applied across it. Calculate the resistance of the lamp. Show your working.

resistance of lamp = ...

13 A p.d. of 1.5 V is applied across a resistor of 5 Ω. Calculate the current in the resistor. Show your working.

current in resistor = ...

14 A p.d. of 1.5 V is applied across two resistors of value 6 Ω and 9 Ω connected in series.

Calculate and show your working for:

a the total resistance

b the current

total resistance = ...

current = ...

c the p.d. across the 6 Ω resistor

d the p.d. across the 9 Ω resistor.

p.d. across 6 Ω resistor = ...

p.d. across 9 Ω resistor = ...

15 Circle the statement that is **false** for a series circuit.

A The current at every point is the same.

B The sum of the p.d.s across each component equals the total p.d. across the supply.

C The combined resistance of the resistors is less than the sum of the individual resistors.

D A complete circuit is needed for there to be a current.

16 Circle the statement that is **incorrect** for an electrical appliance.

A A fuse protects the appliance.

C A fuse should be placed in the neutral wire.

B A fuse should be placed in the live wire.

D The metal case should be earthed.

17 A washing machine is rated at 10 A, 240 V. Calculate and show your working for:

a the power of the washing machine

power = ...

b the energy supplied to the washing machine in 40 minutes.

energy supplied = ...

c The cost of electricity is 10 cents per kWh. Calculate the cost of using the washing machine for 40 minutes.

cost = ...

18 Calculate the size of fuse to be used for a 600 W electric motor connected to a 240 V supply. Show your working.

size of fuse = ...

19 Four appliances are plugged into a power board which is protected by a 13 A fuse and connected to a 240 V supply. The appliances are rated at 1.5 kW, 1.0 kW, 500 W and 300 W.

Circle the **correct** number of appliances that can be turned on together without the fuse blowing. Calculate the current in each appliance and the total current drawn by all four appliances before selecting your answer. Show your working.

A 1   **B** 2   **C** 3   **D** 4

Supplement

20 Explain, with reference to the motion of electrons, why:

a metals are good electrical conductors

....

....

b plastics are good electrical insulators.

....

....

21 A positively charged rod is brought near to a negatively charged electroscope and is then removed without touching it. Identify what happens to the leaf of the electroscope.

Circle the **correct** answer.

The leaf of the electroscope:

A rises more and stays at that deflection

B rises more and then returns to its previous deflection

C falls and stays at that deflection

D falls and returns to its previous deflection.

22 Make a sketch showing the electric field lines:

a between two oppositely charged parallel metal plates

b around a small isolated positive point charge.

[3]

23 The current in a circuit is 5 A. Calculate and show your working for how much charge passes any point in the circuit in:

a 4 seconds

charge = ...

b 2 minutes.

charge = ...

24 Calculate and show your working for the current in a circuit when:

a 30 C of charge flows past a point in a circuit in 15 s

current = ...

b 60 C of charge flows in 1 minute.

current = ...

25 Circle the statement that is **false** for an electric circuit.

A Conventional current flows in a circuit from the positive terminal of a supply to the negative terminal.

B A complete circuit is needed for current to flow.

C Electrons flow in the same direction as the conventional current.

D Electrons flow in the opposite direction to the conventional current.

26 A potential divider circuit is shown below.

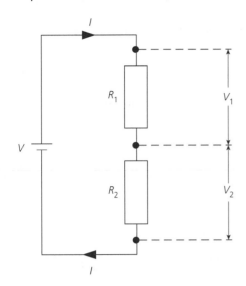

a Circle the **correct** answer.

The ratio $\dfrac{V_1}{V_2}$ of the p.d.s across the two resistors R_1 and R_2 is given by:

A $R_1 + R_2$ **B** $\dfrac{R_1}{R_2}$ **C** $\dfrac{R_2}{R_1}$ **D** $\dfrac{1}{R_1}$

b If $R_1 = 5\,\Omega$, $R_2 = 15\,\Omega$ and $V_1 + V_2 = 10\,V$, calculate **i** V_2 and **ii** V_2. Show your working.

i $V_1 =$ ....   **ii** $V_2 =$ ....

27 The p.d. across a device is 6 V. Calculate and show your working for the energy transferred when:

a a charge of 0.5 C passes through it   **b** the current is 3 A for 20 s.

energy = ...   energy = ...

28 A battery supplies 48 J of energy to a device in 10 seconds. If the p.d. across the device is 12 V, calculate and show your working for:

a the charge flowing through the device in the 10 seconds   **b** the current in the device.

charge = ...   current = ...

29 A number of 1.5 V cells are available.

    a State the number of cells needed and how they should be connected to produce a 6 V battery.

    ....

    b Calculate how much energy the 6 V battery transfers when it drives 2 C around a complete circuit. Show your working.

        energy = ...

30 Give the name and symbol of the unit used to measure the following:

    a charge ....

    b current ....

    c e.m.f. ....

    d potential difference ....

    e resistance ....

31 A p.d. of 6 V is applied across two resistors of value 4 Ω and 12 Ω connected in parallel.

Calculate and show your working for:

    a the total resistance

        total resistance = ...

    b the current from the battery

        current from battery = ...

    c the current in the 4 Ω resistor

        current in 4 Ω resistor = ...

d the current in the 12 Ω resistor.

current in 12 Ω resistor = ...

32 Sketch how the current varies with the p.d. across:

a a resistor of constant resistance   **b** a filament lamp.

33 State how the resistance varies with current direction in:

a a semiconductor diode ....

....

b a filament lamp. ....

34 State how the resistance of a wire changes when:

a its length is doubled ....

b its length is halved ....

c its cross-sectional area is halved ....

d its cross-sectional area is doubled. ....

35 Circle the statement that is **false** for a parallel circuit.

A The current from the source equals the sum of the currents in each branch.

B The p.d. across each component is less than the total p.d. across the supply.

C The combined resistance of resistors is less than that of the resistors individually.

D A complete circuit is needed for there to be a current.

36 Sketch the *I–V* curve for a semiconductor diode.

37 Explain the difference between analogue and digital voltages.

....

....

38 Give **two** examples of:

   a input devices that transfer energy from the surroundings to an electrical circuit

    1 ....

    2 ....

   b output devices that transfer energy from an electrical circuit to the surroundings.

    1 ....

    2 ....

39 A light-dependent resistor (LDR) in series with a resistor is used in a potential divider across a d.c. supply to operate a relay and a lamp, as shown.

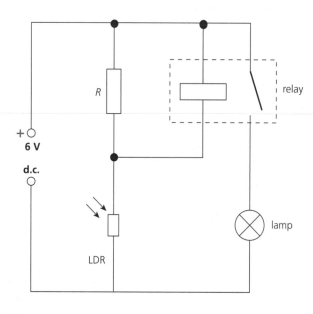

When light falls on the LDR, its resistance and the p.d. across it fall.

a State what happens to the following:

i the p.d. across R, when light falls on the LDR

....

ii the relay, when the p.d. across it reaches its operating p.d.

....

iii the lamp, when the relay reaches its operating p.d. when light falls on the LDR.

....

b Suggest a use for the circuit.

....

Exam-style questions

Core

1 This question is about the hazards of electricity.

a Suggest **two** checks you could make to reduce the risk of receiving an electric shock from an old electrical appliance.

1 ....

2 .... *[2]*

b Explain why the severity of an electric shock is increased by damp conditions.

....

.... *[2]*

c Explain what causes a wire to overheat.

.... *[2]*

d Explain the purpose of a fuse in a circuit.

....

.... *[2]*

[Total: 8]

Supplement

2 Twenty party lights are connected in parallel and operated by a 24 V power supply.

   a Explain why it is preferable to connect party lights in parallel rather than in series.

....

.... **[2]**

   b If each light has a resistance of 80 Ω, calculate:

     i the current in each lamp

current = ... **[2]**

     ii the total current drawn from the power supply

total current from supply = ... **[2]**

     iii the total current drawn from the power supply if one lamp fails.

total current if one lamp fails = ... **[2]**

[Total: 8]

3 A p.d. of 9 V is applied across a potential divider circuit containing a thermistor, R_1, and a fixed resistor, R_2.

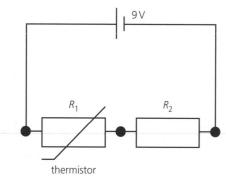

a Calculate the current in the circuit and the values of the p.d.s across R_1 and R_2 when $R_1 = 18\,\Omega$ and $R_2 = 12\,\Omega$.

    i current = ...   *[2]*

    ii p.d. across R_1 = ...   *[3]*

    iii p.d. across R_2 = ...   *[1]*

b Calculate the current in the circuit and the values of the p.d.s across R_1 and R_2 when $R_1 = 33\,\Omega$ and $R_2 = 12\,\Omega$.

    i current = ...   *[2]*

    ii p.d. across R_1 = ...   *[3]*

    iii p.d. across R_2 = ...   *[1]*

c Explain how the p.d. across the thermistor changes as its resistance increases.

    .... *[1]*

[Total: 13]

4 A p.d. of 6 V is connected across a potential divider consisting of a light-dependent resistor (LDR) and a resistor R.

a If the resistance of the LDR decreases, state what happens to:

    i the current in the circuit ....   *[1]*

    ii the p.d. across R ....   *[1]*

    iii the p.d. across the LDR. ....   *[1]*

b If $R = 10\,\Omega$, calculate the resistance of the LDR when the p.d. across it is 3.5 V.

    resistance of LDR = ...   *[4]*

[Total: 7]

4B Electromagnetic effects

Core

1 Circle the **correct** answer.

An e.m.f. is induced in a conductor when it:

A moves parallel to magnetic field lines

B moves across magnetic field lines

C is at rest in a magnetic field

D is heated in the absence of a magnetic field.

2 An electromagnet is constructed by winding a solenoid on a soft iron core. Give **two** ways by which the strength of the electromagnet could be increased.

1 ....

2 ....

3 When a coil is rotated in a magnetic field, an e.m.f. is generated. Give the **three** factors on which the size of the induced e.m.f. depends.

1 ....

2 ....

3 ....

4 Sketch the field around a current-carrying straight wire, showing the direction of the current and of the field lines.

5 A current-carrying straight wire is placed between the poles of a magnet, as shown.

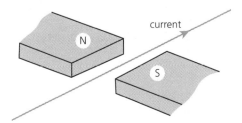

The wire moves downwards when current flows.

State the direction in which the wire moves when:

a the current direction is reversed ....

b the magnetic field direction is reversed. ....

6 Explain why electricity is distributed around the country using a.c. rather than d.c.

....

....

7 Circle the **correct** answer.

The main function of a step-up transformer is to:

A increase power

B increase current

C increase voltage

D decrease resistance of a circuit.

8 240 V a.c. is applied to the primary coil of a transformer which has 600 turns. The secondary coil has 30 turns.

a Calculate the p.d. across the secondary coil. Show your working.

p.d. across secondary coil = ...

b State the type of transformer. ....

Supplement

9 A bar magnet is moved in and out of a coil that is connected to a sensitive centre-zero meter, as shown.

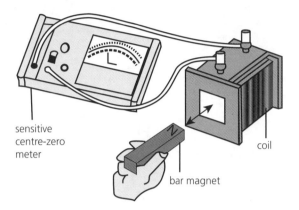

sensitive
centre-zero
meter

coil

bar magnet

The meter needle swings to the left when the magnet is moving towards the coil. State how the needle behaves when the bar magnet:

a is at rest inside the coil ....

b is moving back out of the coil ....

c moves more quickly towards the coil ....

d is at rest and the coil is moved away from the magnet. ....

10 A straight wire is moved upwards through a horizontal magnetic field. Mark on the diagram the direction in which current will flow in the wire if it is connected to a complete circuit.

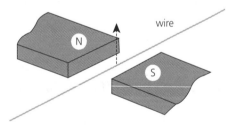

wire

N

S

11 A bar magnet is moved towards a coil of wire connected to a meter, as shown.

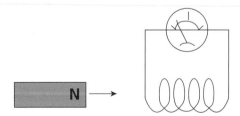

N →

An e.m.f. is induced in the coil and current flows in the circuit making the coil act like a magnet. State the polarity of the end of the coil nearest the magnet when the N pole of the magnet:

a approaches the coil ....

b is pulled back from the coil ....

c is at rest and the coil is moved towards the magnet. ....

12 Electricity transmission lines deliver 80 kW of power to the consumer at a voltage of 400 000 V. The resistance of the transmission lines is 15 Ω.

a Calculate the current in the transmission lines. Show your working.

current = ...

b Calculate the rate of loss of energy from the transmission lines. Show your working.

rate of energy loss = ...

13 The pattern of magnetic field lines in and around a solenoid is shown below.

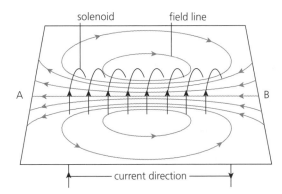

a A solenoid, like a bar magnet, has a N and a S pole.

i Is position A or B nearest to the S pole of the solenoid? ....

ii Explain how the polarity of the solenoid could be reversed. ....

iii In which direction would the N pole of a plotting compass point if placed at A?

....

b State where the magnetic field is:

i strong ....

ii weak. ....

14 The current in the primary coil of a step-down transformer is 0.15 A. The input to the primary coil is 240 V and the output from the secondary coil is 12 V.

Calculate and show your working for the current that can be drawn from the secondary coil if the transformer is:

a 100% efficient in transferring power

secondary coil current = ...

b 90% efficient in transferring power.

secondary coil current = ...

15 A high-speed electron beam enters a region of uniform electric field, produced by a p.d. applied to two parallel metal plates, as shown.

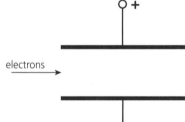

In which direction are the electrons deflected? ....

16 A high-speed electron beam enters a region of uniform magnetic field, as shown.

magnetic field
acting into page

electrons

In which direction are the electrons deflected? ....

Exam-style questions

Core

1 The diagram below represents a relay when it is switched off.

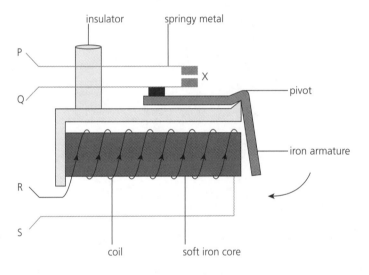

insulator   springy metal

P

X

Q

pivot

iron armature

R

S

coil   soft iron core

a Explain what will cause the relay to switch on. ....
.... *[1]*

b Explain why there is a soft iron core in the relay. ....
.... *[2]*

c State what happens at X when the relay is switched on. ....
.... *[2]*

d Give a reason for using a relay. ....
.... *[2]*

[Total: 7]

2 An electromagnet, P, is used to lift some scrap iron. The table shows the maximum load that can be lifted when there are different currents in the electromagnet.

Current/A	0.5	1.0	1.5	2.0	2.5
Load/N	20	40	60	80	100

a Sketch a load–current graph with current along the x-axis and load along the y-axis.　　**[4]**

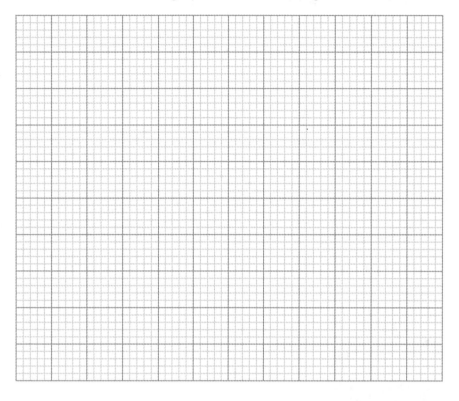

b Describe the shape of your graph. ....　　**[1]**

c Determine the current needed to lift a load of 30 N. ....　　**[1]**

d State the maximum load that could be lifted by a current of 1.25 A. ....　　**[1]**

e A second electromagnet, Q, has the same design as electromagnet P but has twice the number of turns on the coil. State the advantages that electromagnet Q would have over electromagnet P.

.... **[1]**

....

[Total: 8]

Supplement

3 The diagram below shows a simple generator.

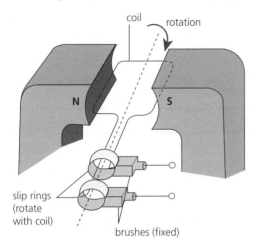

a Explain the function of the slip rings.

.... *[1]*

b Sketch the voltage output of the generator against time for one revolution of the coil.

[3]

c State the position of the coil when the induced voltage is a maximum.

.... *[1]*

d State the type of current there would be in a circuit connected to the brushes.

.... *[1]*

[Total: 6]

4 In a d.c. electric motor, a current-carrying coil attached to a split-ring commutator is mounted between the poles of a magnet. The coil experiences a turning effect.

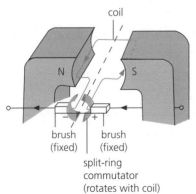

a State the result of the turning effect on the coil.

.... [1]

b State how the size of the turning effect on the coil can be increased.

....

.... [3]

c Describe the effect on the coil of increasing the turning effect.

.... [1]

d Explain the function of the split-ring commutator.

....

.... [2]

[Total: 7]

5 When 250 V a.c. is applied to the primary coil of a step-up transformer, the output from the secondary coil is 1000 V. The current in the primary coil is 0.8 A and there are 100 turns on the primary coil.

a Calculate the number of turns on the secondary coil.

number of turns on secondary coil = ...   [3]

b Calculate the power input to the transformer.

power input = ...   [2]

c Calculate the current in the secondary coil if no energy were lost in the transformer.

current in secondary coil = ...   [3]

d Explain the effect the resistance of the coil windings has on the efficiency of a transformer.

.... [1]

[Total: 9]

5 Nuclear physics

Core

1 Circle the **incorrect** statement.

   A All the positive charge of an atom is concentrated in the nucleus.

   B Most of the mass of an atom is concentrated in the nucleus.

   C Negatively charged electrons orbit the nucleus.

   D The nucleus has a very small size compared with the whole atom.

   E The nucleus consists of protons and neutrons which have a similar mass.

   F Electrons and protons have equal and opposite charge.

   G Electrons and neutrons have the same charge.

   H The mass of an electron is very small compared with the mass of a proton or neutron.

2 The nucleon number of an atom is written as A, the proton number as Z and the neutron number as N.

   a Give the relationship between A, Z and N.

   ....

   b In terms of A, Z and N, how many electrons are there in a neutral atom?

   ....

   c An atom X may be represented in nuclide notation as $^{A}_{Z}X$. In this format, explain how you would represent the following:

     i a helium atom with two neutrons and two protons in its nucleus ....

     ii an electron ....

     iii a neutron. ....

3 Explain the term 'isotopes of an element'.

   ....

   ....

4 Carbon exists as different isotopes.

    a A neutral atom of one isotope of carbon has six protons, six electrons and eight neutrons. Calculate the nucleon number for this isotope.

       nucleon number = ...

    b A second isotope of carbon has a nucleon number of 12. For a neutral atom of this isotope, state the number of:

       i protons ....

       ii neutrons ....

       iii electrons. ....

    c Will the two isotopes have the same chemical properties? ....

5 State which of the following radiations:

α-particles	β-particles	γ-rays

    a causes most ionisation ....

    b causes least ionisation ....

    c is most penetrating ....

    d is least penetrating ....

    e requires a few millimetres of aluminium to stop it ....

    f travels only a few centimetres in air ....

6 If the charge on an electron is denoted as being −1 unit, state the charge on:

    a a proton ....   **d** a β-particle ....

    b a neutron ....   **e** a γ-ray ....

    c an α-particle ....   **f** a helium nucleus. ....

7 State which of the following radiations:

α-particles	β-particles	γ-rays

   a are electrons ....

   b are helium nuclei ....

   c are electromagnetic waves ....

   d are easily deflected by an electric field ....

   e are not deflected by a magnetic field ....

   f have a positive charge ....

   g have a negative charge. ....

8 α, β and γ are ionising radiations.

   a Explain what happens when ionisation of a gas occurs.

   ....

   ....

   b State how ionisation can be detected.

   ....

9 Circle the **incorrect** statement.

   Radioactive decays:

   A occur randomly over space and time   **C** result from unstable nuclei

   B are affected by chemical interactions   **D** produce α-particles, β-particles and γ-rays.

10 Explain the meaning of the terms:

   a half-life

   ....

   ....

   b background radiation.

   ....

   ....

11 Radioactive carbon-14 has a half-life of 5700 years. A 10 g piece of wood found in an archaeological excavation gives a count rate of 80 counts/minute. A 10 g sample of a piece of wood cut recently from a living tree has a count rate of 160 counts/minute.

Estimate the age of the wood taken from the excavation. Show your working.

estimated age of wood = ...

Supplement

12 A beam of α-particles enters a region of uniform magnetic field, as shown.

magnetic field acting into page

α-particles

In which direction would the α-particles initially experience a force? ....

13 In order to monitor paper quality in a factory, a radiation source is placed close to one side of a moving sheet of paper and a Geiger counter is placed on the opposite side.

a Explain what happens to the count-rate on the Geiger counter when the paper passing it:

i becomes thicker ....   **ii** becomes thinner. ....

b Identify the type of radiation source that should be used for this application. Give a reason for your answer.

....

....

14 In a famous experiment α-particles were fired at a thin metal foil. Explain how the behaviour of the α-particles provided evidence for the nuclear model of the atom.

....

....

....

15 The radioactive nuclide iodine-131 decays through β-particle emission as indicated in the equation below.

$$^{131}_{53}I \rightarrow {}^{A}_{Z}Xe + {}^{0}_{-1}e$$

a Calculate the values of A and Z.

i A = ...

ii Z = ...

b Calculate the number of neutrons in the nucleus of an atom of iodine-131.

....

c Determine the number of neutrons in the nucleus of an iodine-131 atom when it decays to xenon.

....

Exam-style questions

Core

1 Give **two** examples of:

a dangers of ionising radiation

1 ....

2 .... *[2]*

b sources of background radiation

1 ....

2 .... *[2]*

c safety precautions that should be taken when handling radioactive materials.

1 ....

2 .... *[2]*

[Total: 6]

Supplement

2 Some large unstable nuclei undergo nuclear fission.

a Explain the term 'nuclear fission'.

....

.... *[2]*

b The equation below represents what may happen when a nucleus of U-235 is struck by and absorbs a neutron.

$$^{235}_{92}U + ^{1}_{0}n \rightarrow ^{A}_{56}Ba + ^{90}_{Z}Kr + ^{1}_{0}n + ^{1}_{0}n$$

Calculate the values of A and Z.

i A = ... *[2]*   **ii** Z = ... *[2]*

c Energy is released in nuclear fission.

i Identify the energy store to which nuclear energy is transferred.

.... *[1]*

ii Describe how this energy is used in a nuclear power station.

.... *[1]*

[Total: 8]

3 In certain conditions some nuclei can undergo nuclear fusion.

a Explain the term 'nuclear fusion'.

....

.... *[2]*

b One of the fusion reactions that occurs in the Sun is given by the following equation.

$$^{3}_{2}He + ^{3}_{2}He \rightarrow ^{A}_{2}X + ^{1}_{1}H + ^{1}_{Z}Y$$

Calculate the values of A and Z.

i A = ... *[2]*   **ii** Z = ... *[2]*

c Identify the nuclides:

i X .... *[1]*   **ii** Y .... *[1]*

d Energy is released in the fusion reactions that take place in the Sun. State how this energy is transferred to us.

.... *[1]*

[Total: 9]

4 A very small amount of radioactive americium dioxide is used in a smoke detector.

The nuclide americium-241 decays to neptunium-237, as shown in the equation below.

$$^{241}_{95}\text{Am} \rightarrow\,^{237}_{93}\text{Np} +\,^{A}_{Z}\text{X}$$

a Calculate the values of A and Z.

i A = ... **[2]**   ii Z = ... **[2]**

b Identify the particle $^{A}_{Z}\text{X}$ emitted in the decay process. .... **[1]**

c State the function that X performs in the smoke detector.

.... **[1]**

[Total: 6]

6 Space physics

Core

1 Circle the statement which is **not true**.

    A The Earth goes around the Sun once each year.

    B The Earth has seasons because its axis is tilted.

    C When the Earth's northern hemisphere is tilted away from the Sun, the night there is shorter than the day.

    D Day and night are due to the Earth spinning on its axis.

2 Circle the statement which is **not true**.

    A The Earth spins on its axis once every 24 hours.

    B The Moon spins on its axis once every 24 hours.

    C The Moon spins on its axis approximately once a month.

    D The Moon orbits the Earth approximately once a month.

3 **a** Identify **two** factors which cause the Earth to have seasons.

      1 ....

      2 ....

    b Explain why it is winter in the northern hemisphere when it is summer in the southern hemisphere.

      ....

      ....

4 Circle the statement which is **incorrect**.

    A There are billions of galaxies in the Universe.

    B A galaxy is a large collection of stars.

    C A light-year is the time it takes for light from the Sun to reach the Earth.

    D The Solar System lies in the Milky Way.

5 Circle the statement which is **incorrect**.

 A The Sun is a star.

 B The Solar System lies in the Andromeda galaxy.

 C There are billions of stars in the Milky Way.

 D A light-year is the distance travelled by light in one year.

6 Choose the distance which matches each of the following statements.

 A 2 million light-years   **C** 4 light-years

 B 100 thousand light-years   **D** 150 million km

 a the distance of the second nearest star to the Earth is more than ....

 b the distance of the nearest galaxy to the Earth is more than ....

 c the distance of the Earth from the Sun is approximately ....

 d the diameter of the Milky Way is approximately ....

Supplement

7 Identify which of the following values is approximately equal to one light-year.

 A $3 \times 10^8 \, \text{m/s}$

 B $9.5 \times 10^{15} \, \text{m}$

 C $10^{16} \, \text{km}$

 D $2.2 \times 10^{-18} \, \text{s}^{-1}$

8 Explain the following facts.

 a A year on Mercury is less than one year on Earth.

 ....

 ....

 b The nights on Mars are very cold.

 ....

 ....

c The four giant outer planets have atmospheres containing hydrogen gas but the Earth does not.

....

....

9 The average distance of Venus from the Sun is 108×10^6 km. Calculate the number of Earth days it takes Venus to complete one orbit if its orbital speed is 35 km/s. Show your working.

number of days = ...

10 Explain how stars are formed. Describe the chemical composition, the influence of gravitational forces, temperature, nuclear reactions and the balancing of forces in the development of a stable star from interstellar clouds of dust and gas.

....

....

....

....

....

....

11 Use the following words to fill in the gaps in the paragraph below.

carbon	collapse	dwarf	energy	expand	fusion	planetary	temperature

The Sun is powered by the release of .... in the nuclear .... of

hydrogen into helium. It will run out of hydrogen in about 5000 million years when its core will

.... into a red giant and its outer layers will .... and cool. In the

core of the red giant, the .... will become high enough for helium to be converted

into .... but eventually all the helium will be used up, the core will collapse again

and a .... nebula with a white .... at its centre will be formed.

12 Give the stages in the life cycle of a star with a mass more than 8 times the mass of the Sun.

.... ⟶ .... ⟶ .... ⟶ ....

13 In the schematic diagram below, showing the life cycle of low mass and high mass stars, fill in the empty boxes A to E by choosing from the following list.

| black hole | supernova | neutron star | planetary nebula | white dwarf |

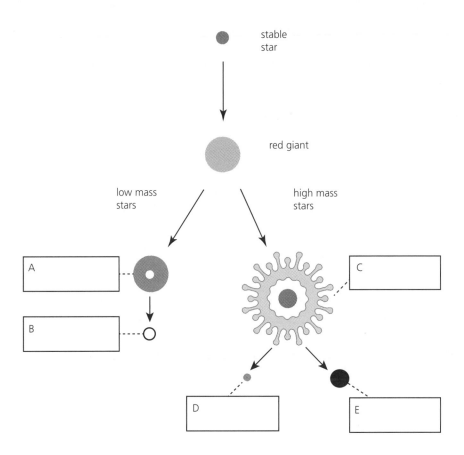

14 Identify which of the following provides evidence in support of the Big Bang theory.

A supernova explosions

B cosmic microwave background radiation

C nuclear fusion of helium into carbon

D black holes

15 State **two** discoveries that support the Big Bang theory of the origin of the Universe.

1 ....

....

2 ....

....

16 Redshift measurements from a distant galaxy show that it is moving away from the Earth at a speed of 350 km/s.

Calculate how far away the galaxy is from the Earth. Show your working.
(Assume $H_0 = 2.2 \times 10^{-18}\,\text{s}^{-1}$.)

distance of the galaxy from Earth = ...

17 Calculate the Hubble constant if a galaxy 1100 million light years from the Earth is receding at 24 000 km/s. Show your working.

Hubble constant = ...

Exam-style questions

Core

1 Daily and seasonal occurrences can be explained by the motion and tilt of the Earth.

   a Explain why the Sun rises in the east and sets in the west each day.

   ....

   .... [2]

   b State when the Sun is at its greatest height above the horizon in the northern hemisphere.

   ....

   .... [2]

   [Total: 4]

2 Astronomers study the spectrum of light from distant stars.

   a State what causes a star to emit radiation in the ultraviolet, visible and infrared parts of the electromagnetic spectrum.

   ....

   .... [2]

   b Explain how this light appears if a distant star is moving away from the Earth.

   ....

   .... [2]

   c Explain how light from distant stars provides evidence that the Universe is expanding.

   ....

   ....

   ....

   ....

   .... [4]

   [Total: 8]

Supplement

3 Stars have several stages in their evolution.

   a Explain how a protostar is formed.

   ....

   .... [1]

b Describe how a star is powered.

....

.... [2]

c Explain when a stable star becomes unstable.

....

.... [1]

d Explain how a red giant is formed.

....

....

.... [3]

[Total: 7]

4 The orbital time of Neptune around the Sun is 165 Earth years and its orbital path is 30 times longer than that of the Earth.

    a Calculate the orbital speed of Neptune relative to the Earth.

        orbital speed = ... [4]

    b State how the speed of travel of Neptune compares with that of the Earth.

.... [1]

[Total: 5]

5 **a** Explain what is meant by 'redshift'.

....

....

.... [3]

    b State Hubble's law.

....

.... [2]

[Total: 5]

6 Redshift measurements show that a galaxy is moving away from the Earth at the speed of 60 000 km/s.

Taking the Hubble constant H_0 as $2.2 \times 10^{-18}\,\text{s}^{-1}$ and 1 light-year as $10^{13}\,\text{km}$, calculate and show your working for:

a how far the galaxy is from the Earth

distance = ... [3]

b how long it has taken light from the galaxy to reach the Earth

time taken = ... [2]

c the time the galaxy and the Milky Way have been travelling apart.

time = ... [5]

d Explain the significance of the time you calculated in **c**.

....

.... [1]

[Total: 11]

Reinforce learning and deepen understanding of the key concepts in this Workbook, which provides additional support for the accompanying Cambridge IGCSE™ Physics Student's Book.

» **Develop understanding and build confidence ahead of assessment:** the Workbook is in syllabus order, topic-by-topic, with each section containing a range of shorter questions to test knowledge, and sections providing exam-style questions.

» **Differentiated content:** both Core and Supplement content is clearly flagged with differentiated questions.

» **Provide extra practice and self-assessment:** each Workbook is intended to be used by students for practice and homework. Once completed, it can be kept and used for revision.

Use with *Cambridge IGCSE™ Physics Student's Book* **Fourth Edition** 9781398310544

For over 25 years we have been trusted by Cambridge schools around the world to provide quality support for teaching and learning. For this reason we have been selected by Cambridge Assessment International Education as an official publisher of endorsed material for their syllabuses.

This resource is endorsed by Cambridge Assessment International Education

✓ Provides learner support for the Cambridge IGCSE™ and IGCSE™ (9-1) Physics syllabuses (0625/0972) for examination from 2023

✓ Has passed Cambridge International's rigorous quality-assurance process

✓ Developed by subject experts

✓ For Cambridge schools worldwide

HODDER EDUCATION
e: education@hachette.co.uk
w: hoddereducation.com

ISBN 978-1-398-31057-5

9 781398 310575

MIX
Paper from responsible sources
FSC™ C104740